Life Cycles

Level 11 – Lime

Helpful Hints for Reading at Home

The graphemes (written letters) and phonemes (units of sound) used throughout this series are aligned with Letters and Sounds. This offers a consistent approach to learning whether reading at home or in the classroom.

HERE ARE SOME COMMON WORDS THAT YOUR CHILD MIGHT FIND TRICKY:

water	where	would	know	thought	through	couldn't
laughed	eyes	once	we're	school	can't	our

TOP TIPS FOR HELPING YOUR CHILD TO READ:

- Encourage your child to read aloud as well as silently to themselves.
- Allow your child time to absorb the text and make comments.
- Ask simple questions about the text to assess understanding.
- Encourage your child to clarify the meaning of new vocabulary.

This book focuses on developing independence, fluency and comprehension. It is a Lime level 11 book band.

©2022 **BookLife Publishing Ltd.**
King's Lynn, Norfolk, PE30 4LS, UK

ISBN 978-1-80155-820-4

All rights reserved. Printed in Poland.
A catalogue record for this book is available from the British Library.

Life Cycles
Written by Louise Nelson
Adapted by Hermione Redshaw
Designed by Drue Rintoul

Image Credits Images are courtesy of Shutterstock.com. With thanks to Getty Images, Thinkstock Photo and iStockphoto. Cover – Samuel Borges Photography, Nadzin, Rvector, ExpressVectors, robuart. p4–5 – Ermolaeva Olga 84, Monkey Business Images. p6–7 – Rebecca Schreiner. p8–9 – encierro, showcake. p10–11 – meunierd, Josephine Julian. p12–13 – cynoclub, Ginny Thomas. p14–15 – worldswildlifewonders, Cathy Keifer. p16–17 – Martin Mecnarowski, Alina Boldina. p18–19 – Nilesh Bhadla, Elvira Tursynbayeva. p20–21 – IntoTheWorld, Sari ONeal.

Contents

- Page 4   Living Things
- Page 6   The Journey through Life
- Page 8   Seeds
- Page 10 Eggs
- Page 12 Pregnancy
- Page 14 Offspring
- Page 16 Juveniles
- Page 18 Adulthood
- Page 20 Believe It or Not!
- Page 21 Activity
- Page 22 Index
- Page 23 Questions

Living Things

You are a living thing. So are animals and plants. A dry leaf on the ground was once part of a living thing, but it is no longer living. Now, it is dead. Every living thing follows the same path through life. This is called the life cycle.

All lives have a beginning, a middle, and an end. Living things start off small and young. They grow and change, and have offspring of their own. All living things need to make more of themselves so that their species carries on. They will grow old. Eventually, all living things will die. However, the life cycle continues with their offspring.

The Journey through Life

All living things have a life span. This is the usual amount of time they will probably live. Life spans can be very short or very long. It is thought that the oldest human was Jeanne Louise Calment, who was born in 1875 and lived to the age of 122 years old.

Living things take the same journey through life. Some life cycles might only last for a few hours or days, while others last for hundreds of years. However, the immortal jellyfish is very special, because it doesn't get old. If it gets attacked or injured, it can turn back into a baby and start again.

Seeds

Some plants start life as seeds. Seeds are made inside flowers, fruit or cones. Seeds need to leave the parent plant and find a good place to grow into new plants. They can be carried away from the plant by wind, water or animals. Seeds have coats that protect them from extreme temperatures, injury and parasites.

Seeds need water, air and the right temperature to grow. When a seed's conditions are right, the seed will put out tiny roots and a small stalk called a shoot. The shoot will grow small leaves. This is called germination. Plants all do this differently, but the basic process is the same.

Eggs

Many animals, including birds, reptiles and insects, lay eggs. First, the female will find a mate who will fertilise her eggs. She will then lay her eggs on the ground, with the offspring growing inside them. Fish eggs are often soft and jelly-like. Bird eggs have a hard shell. Reptile eggs can have a soft shell.

Inside the egg, the new offspring will grow. An egg gives the offspring all the food it needs. Some animals will stay with their eggs to protect them and keep them warm. Other animals will lay their eggs and then leave, without looking after them at all. Some animals lay millions of eggs in a clutch – others lay just one.

Pregnancy

Mammals don't usually lay eggs. The offspring of mammals grow inside the mother. Her body gives it all the food it needs. The amount of time a mammal is pregnant is called gestation, and it is different for each species. Usually, smaller animals are pregnant for short periods of time, and bigger animals are pregnant for longer.

The mother will give birth to her offspring. This means she will push them out of her body. Some mammals have just one or two offspring, while others can have more. Some offspring are small and helpless when they are born. They need care and time to grow. Other offspring can stand and run soon after being born.

Offspring

Some animal offspring quickly look like smaller versions of their parents. These small animals will be at risk of being hunted by predators. Some parents protect their offspring, carrying them around and feeding them. Other babies have to look after themselves. These animals often are very good at hiding from predators.

Some animal offspring don't look anything like their parents when they are born or hatch from their eggs. These creatures have a special trick: metamorphosis. They will be born in one body and then, when they have eaten enough to grow, they will change their bodies completely to become like the parent.

Butterflies start off as caterpillars before metamorphosis.

Juveniles

Young birds leave the nest and learn to fly. At this stage, the bird is called a fledgling. Young mammals will stay with their mothers and drink their milk until they are ready to feed themselves. When they stop this, they are known as juveniles. Juvenile animals will learn all the skills they need to survive from their parents.

After the egg stage, many fish also become larvae. Larvae still have the yolk from their egg attached. They grow into fry when they can eat on their own.

If conditions are right, juvenile plants can rapidly grow and will soon reach maturity. Young trees are called saplings. Usually, saplings are not ready to make fruit or have flowers.

Adulthood

When an animal or plant is ready to reproduce and have offspring of its own, it is an adult. Some living things will continue to grow throughout their lives, while others will stop growing when they reach adulthood. Adults will be completely independent. This means they can feed themselves, protect themselves, and know what they have to do.

Adult animals look for a mate so they can make offspring of their own. Some animals only meet up with others to mate, and then they leave. Others will keep the same mate for life. Some adult plants attract bees and other insects by producing flowers. The insects who feed on the flowers help pollinate the plant.

Believe It or Not!

Kangaroos are only pregnant for around four weeks. The baby joey is born very small and underdeveloped. The tiny baby crawls into a pouch on its mother and grows there. At around four months, the joey can leave the pouch for short trips, but it doesn't leave the pouch for good until ten months.

Animals that have this special pouch are called marsupials.

Birds and some reptiles build nests to keep their eggs safe. Some build amazing nests with decorations. The male bower bird must build an impressive nest to attract a female, so he decorates his nest with brightly coloured things.

Female black widow spiders can be very grumpy partners! The female is bigger than the male, and sometimes she eats him!

Index

babies 7, 14, 20
birds 10, 16, 21
life cycles 4–5, 7
life spans 6
mammals 12–13, 16

metamorphosis 15
parasites 8
plants 4, 8–9, 17–19
saplings 17
species 5, 12

How to Use an Index

An index helps us to find information in a book. Each word has a set of page numbers. These page numbers are where you can find information about that word.

Page numbers

Example: balloons 5, <u>8–10</u>, 19

Important word

This means page 8, page 10, and all the pages in between. Here, it means pages 8, 9 and 10.

Questions

1. What was the name of the animal that doesn't get old?

2. How old was the oldest human being?

3. Name an animal that goes through metamorphosis.

4. Can you use the contents page to find out what juvenile birds are called?

5. Can you use the index page to find saplings in the book?